ON ELECTRICITY

ON ELECTRICITY

The Address On the Occasion of the Commemoration of the Introduction of Niagara Falls Power In Buffalo At the Ellicot Club, January 12, 1897

NIKOLA TESLA

WILDSIDE PRESS

ON ELECTRICITY

This edition published in 2006 by Wildside Press, LLC.
www.wildsidepress.com

I have scarcely had courage enough to address an audience on a few unavoidable occasions, and the experience of this evening, even as disconnected from the cause of our meeting, is quite novel to me. Although in those few instances, of which I have retained agreeable memory, my words have met with a generous reception, I never deceived myself, and knew quite well that my success was not due to any excellency in the rhetorical or demonstrative art. Nevertheless, my sense of duty to respond to the request with which I was honored a few days ago was strong enough to overcome my very grave apprehensions in regard to my ability of doing justice to the topic assigned to me. It is true, at times—even now, as I speak—my mind feels full of the subject, but I know that, as soon as I shall attempt expression, the fugitive conceptions will vanish, and I shall experience certain well known sensations of abandonment, chill and silence. I can see already your disappointed countenances and can read in them the painful regret of the mistake in your choice.

These remarks, gentlemen, are not made with selfish desire of winning your kindness and indulgence on my shortcomings, but with the honest intention of offering you an apology for your disappointment. Nor are they made—as you might be disposed to think—in that playful spirit which, to the enjoyment of the listeners is often displayed by belated speakers. On the contrary, I am deeply earnest in my wish that I were capable of having the fire of eloquence kindled in me, that I might dwell in adequate terms on this fascinating science of electricity, on the marvelous development which electrical annals have recorded and which, as one of the speakers justly remarked, stamp this age as the Electrical Age, and particularly on the great event we are commemorating this day. Unfortunately, this my desire must remain unfulfilled, but I am hopeful that in my formless and incomplete statements, among the few ideas and facts I shall mention there may be something of interest and usefulness, something befitting this unique occasion.

Gentlemen, there are a number of features clearly discernible in, and characteristic of, human intellectual progress in more recent times—features which afford great comfort to the minds of all those who have really at heart the advancement and welfare of mankind.

First of all the inquiry, by the aid of the microscope and electrical instruments of precision, into the nature of our organs and senses, and particularly of those through which we commune directly with the outside world and through which knowledge is conveyed to our minds, has revealed their exact construction and mode of action, which is in conformity with simple and well established physical principles and laws. Hence the observations we make and the facts we ascertain by their help are real facts and observations, and our knowledge is true knowledge. To illustrate:

Our knowledge of form, for instance, is dependent upon the positive fact that light propagates in straight lines, and, owing to this, the image formed by a lens is exactly similar to the object seen. Indeed, my thoughts in such fields and directions have led me to the conclusion that most all human knowledge is based on this simple truth, since practically every idea or conception—and therefore all knowledge—presupposes visual impressions. But if light would not propagate in accordance with the law mentioned, but in conformity with any other law which we might presently conceive, whereby not only the image might not bear any likeness to the object seen, but even the images of the same object at different times or distances might not resemble each other, then our knowledge of form would be very defective, for then we might see, for example, a three-cornered figure as a six or twelve-cornered one. With the clear understanding of the mechanism and mode of action of our organs, we remove all doubts as to the reality and truth of the impressions received from the outside, and thus we bar out—forever, we may hope—that unhealthy speculation and skepticism into which formerly even strong minds were apt to fall.

Let me tell you of another comforting feature. The progress in a measured time is nowadays more rapid and greater than it ever was before. This is quite in accordance with the fundamental law of motion, which commands acceleration and increase of momentum or accumulation of energy under the action of a continuously acting force and tendency, and is the more true as every advance weakens the elements tending to produce friction and retardation. For, after all, what is progress, or—more correctly—development, or evolution, if not a movement, infinitely complex and often unscrutinizable, it is true, but nevertheless exactly determined in quantity as well as in quality of motion by the physical conditions and laws governing? This feature of more recent development is best shown in the rapid merging together of the various arts and sciences by the obliteration of the hard and fast lines of separation, of borders, some of which only a few years ago seemed unsurpassable, and which, like veritable Chinese walls, surrounded every department of inquiry and barred progress. A sense of connectedness of the various apparently widely different forces and phenomena we observe is taking possession of our minds, a sense of deeper understanding of nature as a whole, which, though not yet quite clear and defined, is keen enough to inspire us with the confidence of vast realizations in the near future.

But these features chiefly interest the scientific man, the thinker and reasoner. There is another feature which affords us still more satisfaction and enjoyment, and which is of still more universal interest, chiefly because of its bearing upon the welfare of mankind. Gentlemen, there is an influence which is getting strong and stronger day by day, which shows itself more and more in all departments of human activity, and influence most fruitful and beneficial—the influence of the artist. It was a happy day for the mass of humanity when the artist felt the desire of becoming a physician, an electrician, an engineer or mechanician or—whatnot—a mathematician or a financier; for it was he who wrought all these wonders and grandeur we are witnessing. It was he who abolished that small, pedantic, narrow-grooved school teaching which made of an aspiring student a galley-slave, and he who allowed freedom in the choice of subject of study according to one's pleasure and inclination, and so facilitated development.

Some, who delight in the exercise of the powers of criticism, call this an asymmetrical development, a degeneration or departure from the normal, or even a degradation of the race. But they are mistaken. This is a welcome state of things, a blessing, a wise subdivision of labors, the establishment of conditions most favorable to progress. Let one concentrate all his energies in one single great effort, let him perceive a single truth, even though he be consumed by the sacred fire, then millions of less gifted men can easily follow. Therefore it is not as much quantity as quality of work which determines the magnitude of the progress.

It was the artist, too, who awakened that broad philanthropic spirit which, even in old ages, shone in the teachings of noble reformers and philosophers, that spirit which makes men in all departments and positions work not as much for any material benefit or compensation—though reason may command this also—but chiefly for the sake of success, for the pleasure there is in achieving it and for the good they might be able to do thereby to their fellow-men. Through his influence types of men are now pressing forward, impelled by a deep love for their study, men who are doing wonders in their respective branches, whose chief aim and enjoyment is the acquisition and spread of knowledge, men who look far above earthly things, whose banner is Excelsior! Gentlemen, let us honor the artist; let us thank him, let us drink his health!

Now, in all these enjoyable and elevating features which characterize modern intellectual development, electricity, the expansion of the science of electricity, has been a most potent factor. Electrical science has revealed to us the true nature of light, has provided us with innumerable appliances and instruments of precision, and has thereby vastly added to the exactness of our knowledge. Electrical science has disclosed to us the more intimate relation existing between widely different forces and phenomena and has thus led us to a more complete comprehension of Nature and its many manifestations to our senses. Electrical science, too, by its fascination, by its promises of immense realizations, of wonderful possibilities chiefly in humanitarian respects, has attracted the attention and enlisted the energies of the artist; for where is there a field in which his God-given powers would be of a greater benefit to his fellow-men than this unexplored, almost virgin, region, where, like in a silent forest, a thousand voices respond to every call?With these comforting features, with these cheering prospects, we need not look with any feeling of incertitude or apprehension into the future. There are pessimistic men, who, with anxious faces, continuously whisper in your ear that the nations are secretly arming—arming to the teeth; that they are going to pounce upon each other at a given signal and destroy themselves; that they are all trying to outdo that victorious, great, wonderful German army, against which there is no resistance, for every German has the discipline in his very blood—every German is a soldier, But these men are in error. Look only at our recent experience with the British in that Venezuela difficulty. Two other nations might have crashed together, but not the Anglo-Saxons; they are too far ahead. The men who tell you this are ignoring forces which are continually at work, silently but resistlessly—forces which say Peace!

There is the genuine artist, who inspires us with higher and nobler sentiments, and makes us abhor strife and carnage. There is the engineer, who bridges gulfs and chasms, and facilitates contact and equalization of the heterogeneous masses of humanity. There is the mechanic, who comes with his beautiful time and energy-saving appliances, who perfects his flying machine, not to drop a bag of dynamite on a city or vessel, but to facilitate transport and travel. There, again, is the chemist, who opens new resources and makes existence more pleasant and secure; and there is the electrician, who sends his messages of peace to all parts of the globe. The time will not be long in coming when those men who are turning their ingenuity to inventing quick-firing guns, torpedoes and other implements of destruction—all the while assuring you that it is for the love and good of humanity—will find no takers for their odious tools, and will realize that, had they used their inventive talent in other directions; they might have reaped a far better reward than the sestertia received. And then, and none too soon the cry will be echoed everywhere. Brethren, stop these high-handed methods of the strong, these remnants of barbarism so inimical to progress! Give that valiant warrior opportunities for displaying a more commendable courage than that he shows when, intoxicated with victory, he rushes to the destruction of his fellow-men. Let him toil day and night with a small chance of achieving and yet be unflinching; let him challenge the dangers of exploring the heights of the air and the depths of the sea; let him brave the dread of the plague, the heat of the tropic desert and the ice of the polar region. Turn your energies to warding off the common enemies and danger, the perils that are all around you, that threaten you in the air you breathe, in the water you drink, in the food you consume. It is not strange, is it not shame, that we, beings in the highest state of development in this our world, beings with such immense powers of thought and action, we, the masters of the globe, should be absolutely at the mercy of our unseen foes, that we should not know whether a swallow of food or drink brings joy and life or pain and destruction to us! In this most modern and sensible warfare, in which the bacteriologist leads, the services electricity will render will prove invaluable. The economical production of high-frequency currents, which is now an accomplished fact, enables us to generate easily and in large quantities ozone for the disinfection of the water and the air, while certain novel radiations recently discovered give hope of finding effective remedies against ills of microbic origin, which have heretofore withstood all efforts of the physician. But let me turn to a more pleasant theme.

I have referred to the merging together of the various sciences or departments of research, and to a certain perception of intimate connection between the manifold and apparently different forces and phenomena. Already we know, chiefly through the efforts of a bold pioneer, that light, radiant heat, electrical and magnetic actions are closely related, not to say identical. The chemist professes that the effects of combination and separation of bodies he observes are due to electrical forces, and the physician and physiologist will tell you that even life's progress is electrical. Thus electrical science has gained a universal meaning, and with right this age can claim the name "Age of Electricity."

I wish much to tell you on this occasion—I may say I actually burn for desire of telling you—what electricity really is, but I have very strong reasons, which my coworkers will best appreciate, to follow a precedent established by a great and venerable philosopher, and I shall not dwell on this purely scientific aspect of electricity.

There is another reason for the claim which I have before stated which is even more potent than the former, and that is the immense development in all electrical branches in more recent years and its influence upon other departments of science and industry. To illustrate this influence I only need to refer to the steam or gas engine. For more than half a century the steam engine has served the innumerable wants of man. The work it was called to perform was of such variety and the conditions in each case were so different that, of necessity, a great many types of engines have resulted. In the vast majority of cases the problem put before the engineer was not as it should have been, the broad one of converting the greatest possible amount of heat energy into mechanical power, but it was rather the specific problem of obtaining the mechanical power in such form as to be best suitable for general use. As the reciprocating motion of the piston was not convenient for practical purposes, except in very few instances, the piston was connected to a crank, and thus rotating motions was obtained, which was more suitable and preferable, though it involved numerous disadvantages incident to the crude and wasteful means employed. But until quite recently there were at the disposal of the engineer, for the transformation and transmission of the motion of the piston, no better means than rigid mechanical connections. The past few years have brought forcibly to the attention of the builder the electric motor, with its ideal features. Here was a mode of transmitting mechanical motion simpler by far, and also much more economical. Had this mode been perfected earlier, there can be no doubt that, of the many different types of engine, the majority would not exist, for just as soon as an engine was coupled with an electric generator a type was produced capable of almost universal use. From this moment on there was no necessity to endeavor to perfect engines of special designs capable of doing special kinds of work. The engineer's task became now to concentrate all his efforts upon one type, to perfect one kind of engine—the best; the universal, the engine of the immediate future; namely, the one which is best suitable for the generation of electricity. The first efforts in this direction gave a strong impetus to the development of the reciprocating high speed engine, and also to the turbine, which latter was a type of engine of very limited practical usefulness, but became, to a certain extent, valuable in connection with the electric generator and motor. Still, even the former engine, though improved in many particulars, is not radically changed, and even now has the same objectionable features and limitations. To do away with these as much as possible, a new type of engine is being perfected in which more favorable conditions for economy are maintained, which expands the working fluid with utmost rapidity and loses little heat on the walls, an engine stripped of all usual regulating mechanism—packings, oilers and other appendages—and forming part of an electric generator; and in this type, I may say, I have implicit faith.

The gas or explosive engine has been likewise profoundly affected by the commercial introduction of electric light and power, particularly in quite recent years. The engineer is turning his energies more and more in this direction, being attracted by the prospect of obtaining a higher thermodynamic efficiency. Much larger engines are now being built, the construction is constantly improved, and a novel type of engine, best suitable for the generation of electricity, is being rapidly evolved.

There are many other lines of manufacture and industry in which the influence of electrical development has been even more powerfully felt. So, for instance, the manufacture of a great variety of articles of metal, and especially of chemical products. The welding of metals by electricity, though involving a wasteful process, has, nevertheless, been accepted as a legitimate art, while the manufacture of metal sheet, seamless tubes and the like affords promise of much improvement. We are coming gradually, but surely, to the fusion of bodies and reduction of all kinds of ores—even of iron ores—by the use of electricity, and in each of these departments great realizations are probable. Again, the economical conversion of ordinary currents of supply into high-frequency currents opens up new possibilities, such as the combination of the atmospheric nitrogen and the production of its compounds; for instance, ammonia and nitric acid, and their salts, by novel processes.

The high-frequency currents also bring us to the realization of a more economical system of lighting; namely by means of phosphorescent bulbs or tubes, and enable us to produce with these appliances light of practically any candle-power. Following other developments in purely electrical lines, we have all rejoiced in observing the rapid strides made, which, in quite recent years, have been beyond our most sanguine expectations. To enumerate the many advances recorded is a subject for the reviewer, but I can not pass without mentioning the beautiful discoveries of Lenart and Roentgen, particularly the latter, which have found such a powerful response throughout the scientific world that they have made us forget, for a time, the great achievement of Linde in Germany, who has effected the liquefaction of air on an industrial scale by a process of continuous cooling: the discovery of argon by Lord Rayleigh and Professor Ramsay, and the splendid pioneer work of Professor Dewar in the field of low temperature research. The fact that the United States have contributed a very liberal share to this prodigious progress must afford to all of us great satisfaction. While honoring the workers in other countries and all those who, by profession or inclination, are devoting themselves to strictly scientific pursuits, we have particular reasons to mention with gratitude the names of those who have so much contributed to this marvelous development of electrical industry in this country. Bell, who, by his admirable invention enabling us to transmit speech to great distances, has profoundly affected our commercial and social relations, and even our very mode of life; Edison, who, had he not done anything else beyond his early work in incandescent lighting, would have proved himself one of the greatest benefactors of the age; Westinghouse, the founder of the commercial alternating system; Brush, the great pioneer of arc lighting; Thomson, who gave us the first practical welding machine, and who, with keen sense, contributed very materially to the development of a number of scientific and industrial branches; Weston, who once led the world in dynamo design, and now leads in the construction of electric instruments; Sprague, who, with rare energy, mastered the problem and insured the success of practical electrical railroading; Acheson, Hall, Willson and others, who are creating new and revolutionizing industries here under our very eyes at Niagara. Nor is the work of these gifted men nearly finished at this hour. Much more is still to come, form fortunately, most of them are still full of enthusiasm and vigor. All of these men and many more are untiringly at work investigating new regions and opening up unsuspected and promising fields. Weekly, if not daily, we learn through the journals of a new advance into some unexplored region, where at every step success beckons friendly, and leads the toiler on to hard and harder tasks.

But among all these many departments of research, these many branches of industry, new and old, which are being rapidly expanded, there is one dominating all others in importance—one which is of the greatest significance for the comfort and welfare, not to say for the existence, of mankind, and that is the electrical transmission of power. And in this most important of all fields, gentlemen, long afterwards, when time will have placed the events in their proper perspective, and assigned men to their deserved places, the great event we are commemorating today will stand out as designating a new and glorious epoch in the history of humanity—an epoch grander than that marked by the advent of the steam engine. We have many a monument of past ages: we have the palaces and pyramids, the temples of the Greek and the cathedrals of Christendom. In them is exemplified the power of men, the greatness of nations, the love of art and religious devotion. But that monument at Niagara has something of its own, more in accord with our present thoughts and tendencies. It is a monument worthy of our scientific age, a true monument of enlightenment and of peace. It signifies the subjugation of natural forces to the service of man, the discontinuance of barbarous methods, the relieving of millions from want and suffering. No matter what we attempt to do, no matter to what fields we turn our efforts, we are dependent on power. Our economists may propose more economical systems of administration and utilization of resources, our legislators may make wiser laws and treaties, it matters little; that kind of help can be only temporary. If we want to reduce poverty and misery, if we want to give to every deserving individual what is needed for a safe existence of an intelligent being, we want to provide more machinery, more power. Power is our mainstay, the primary source of our many-sided energies.

With sufficient power at our disposal we can satisfy most of our wants and offer a guaranty for safe and comfortable existence to all, except perhaps to those who are the greatest criminals of all—the voluntarily idle.

The development and wealth of a city, the success of a nation, the progress of the whole human race, is regulated by the power available. Think of the victorious march of the British, the like of which history has never recorded. Apart from the qualities of the race, which have been of great moment, they own the conquest of the world to—coal. For with coal they produce their iron; coal furnishes them light and heat; coal drives the wheels of their immense manufacturing establishments, and coal propels their conquering fleets. But the stores are being more and more exhausted; the labor is getting dearer and dearer, and the demand is continuously increasing. It must be clear to every one that soon some new source of power supply must be opened up, or that at least the present methods must be materially improved. A great deal is expected from a more economical utilization of the stored energy of the carbon in a battery; but while the attainment of such a result would be hailed as a great achievement; it would not be as much of an advance towards the ultimate and permanent method of obtaining power as some engineers seem to believe. By reasons both of economy and convenience we are driven to the general adoption of a system of energy supply from central stations, and for such purposes the beauties of the mechanical generation of electricity can not be exaggerated. The advantages of this universally accepted method are certainly so great that the probability of replacing the engine dynamos by batteries is, in my opinion, a remote one, the more so as the high-pressure steam engine and gas engine give promise of a considerably more economical thermodynamic conversion. Even if we had this day such an economical coal battery, its introduction in central stations would by no means be assured, as its use would entail many inconveniences and drawbacks. Very likely the carbon could not be burned in its natural form as in a boiler, but would have to be specially prepared to secure uniformity in the current generation. There would be a great many cells needed to make up the electro-motive force usually required. The process of cleaning and renewal, the handling of nasty fluids and gases and the great space necessary for so many batteries would make it difficult, if not commercially unprofitable, to operate such a plant in a city or densely populated district. Again if the station be erected in the outskirts, the conversion by rotating transformers or otherwise would be a serious and unavoidable drawback. Furthermore, the regulating appliances and other accessories which would have to be provided would probably make the plant fully as much, if not more, complicated than the present. We might, of course, place the batteries at or near the coal mine, and from there transmit the energy to distant points in the form of high-tension alternating currents obtained from rotating transformers, but even in this most favorable case the process would be a barbarous one, certainly more so than the present, as it would still involve the consumption of material, while at the same time it would restrict the engineer and mechanic in the exercise of their beautiful art. As to the energy supply in small isolated places as dwellings, I have placed my confidence in the development of a light storage battery, involving the use of chemicals manufactured by cheap water power, such as some carbide or oxygen-hydrogen cell.

But we shall not satisfy ourselves simply with improving steam and explosive engines or inventing new batteries; we have something much better to work for, a greater task to fulfill. We have to evolve means for obtaining energy from stores which are forever inexhaustible, to perfect methods which do not imply consumption and waste of any material whatever. Upon this great possibility, which I have long ago recognized, upon this great problem, the practical solution of which means so much for humanity, I have myself concentrated my efforts since a number of years, and a few happy ideas which came to me have inspired me to attempt the most difficult, and given me strength and courage in adversity. Nearly six years ago my confidence had become strong enough to prompt me to an expression of hope in the ultimate solution of this all dominating problem. I have made progress since, and have passed the stage of mere conviction such as is derived from a diligent study of known facts, conclusions and calculations. I now feel sure that the realization of that idea is not far off. But precisely for this reason I feel impelled to point out here an important fact, which I hope will be remembered. Having examined for a long time the possibilities of the development I refer to, namely, that of the operation of engines on any point of the earth by the energy of the medium, I find that even under the theoretically best conditions such a method of obtaining power can not equal in economy, simplicity and many other features the present method, involving a conversion fo the mechanical energy of running water into electrical energy and the transmission of the latter in the form of currents of very high tension to great distances. Provided, therefore, that we can avail ourselves of currents of sufficiently high tension, a waterfall affords us the most advantageous means of getting power from the sun sufficient for all our wants, and this recognition has impressed me strongly with the future importance of the water power, not so much because of its commercial value, though it may be very great, but chiefly because of its bearing upon our safety and welfare. I am glad to say that also in this latter direction my efforts have not been unsuccessful, for I have devised means which will allow us the use in power transmission of electromotive forces much higher than those practicable with ordinary apparatus. In fact, progress in this field has given me fresh hope that I shall see the fulfillment of one of my fondest dreams; namely, the transmission of power from station to station without the employment of any connecting wire. Still, whatever method of transmission be ultimately adopted, nearness to the source of power will remain an important advantage.

Gentlemen, some of the ideas I have expressed may appear to many of you hardly realizable; nevertheless, they are the result of long-continued thought and work. You would judge them ore justly if you would have devoted your life to them, as I have done. With ideas it is like with dizzy heights you climb: At first they cause you discomfort and you are anxious to get down, distrustful of your own powers; but soon the remoteness of the turmoil of life and the inspiring influence of the altitude calm your blood; your step gets firm and sure and you begin to look—for dizzier heights. I have attempted to speak to you on "Electricity," its development and influence, but I fear that I have done it much like a boy who tries to draw a likeness with a few straight lines. But I have endeavored to bring out one feature, to speak to you in one strain which I felt sure would find response in the hearts of all of you, the only one worthy of this occasion—the humanitarian. In the great enterprise at Niagara we see not only a bold engineering and commercial feat, but far more, a giant stride in the right direction as indicated both by exact science and philanthropy. Its success is a signal for the utilization of water powers all over the world, and its influence upon industrial development is incalculable. We must all rejoice in the great achievement and congratulate the intrepid pioneers who have joined their efforts and means to bring it about. It is a pleasure to learn of the friendly attitude of the citizens of Buffalo and of the encouragement given to the enterprise by the Canadian authorities. We shall hope that other cities, like Rochester on this side and Hamilton and Toronto in Canada, will soon follow Buffalo's lead. This fortunate city herself is to be congratulated. With resources now unequaled, with commercial facilities and advantages such as few cities in the world possess, and with the enthusiasm and progressive spirit of its citizens, it is sure to become one of the greatest industrial centers of the globe.

Made in United States
North Haven, CT
04 November 2026

21771749R00016